Tangram Treasury, Book A

by

Jan Fair

Cuisenaire Company of America, Inc.
10 Bank Street. P.O. Box 5026
White Plains, NY 10602-5026

To my children, Stephanie and Eric,
and to all their friends and classmates at Alvin School
who helped pilot-test these activities.

Cover and Illustrations by Rita Formaro

Copyright© 1987
Cuisenaire Company of America, Inc.
10 Bank Street, PO Box 5026, White Plains, N.Y., 10602-5026

ISBN 0-914040-53-7

Printed in U.S.A.

2 3 4 5 6 7 8 9 10 - BK - 98 97

CONTENTS

INTRODUCTION...4

CLASSROOM LESSON..6
Introducing The Tangram

STUDENT WORKSHEETS...7
The Seven Tangram Pieces; Parallelogram Flip; Flip and Fit; What's
My Shape?; Flying Shapes; A Dog-Gone Puzzle; Tangram T-Party;
Covered Baby Carriage; Tangram Lady; Spider Art; hid-Tan Picture

GAMES AND SPINNERS..18
Cover Up; Geometry Spin

STUDENT WORKSHEETS...20
Modern Art; Kite Flying Shapes; Trace A Racer; Don't Flip! Rotate!;
Triangle Pinwheel; Flower Rotations

PROJECT..26
Transformational Tangram Flowers; Tangram Flowers

STUDENT WORKSHEETS...28
Squares In Squares; Triangles In Triangles; A Fishy Cover-Up; Two-
Way Triangles; Cover It Well; Funny Bunny

LEARNING CENTERS...34
Self-Checking Task Cards; Worksheet "Instant" Centers; Independent
Art Projects

TASK CARDS..35
Cat; Dog; Chicken; Duck; Trees; Barn; Tangram Farm

bulleTAN BOARDS...42
Letters And Borders; Tangram Farm; Tangram Art

FAMILY INVOLVEMENT..44
Repeated Success; A Student-Made Activity; Family Games

TAN-gram, TANGRAM PATTERN, and TANGRAM CERTIFICATE................45

ANSWERS..47

INTRODUCTION

The tangram is an old Chinese puzzle game consisting of seven pieces called 'tans'. It's known in China as "the seven clever pieces". And, indeed it is!

TAN'S BACKGROUND

No one knows for sure when this puzzle was first developed. Sam Lloyd, the famous American puzzlist, said it was some 4000 years ago. Others say it was more like 200 years ago. I say, it doesn't matter. What's important, is it's here now!

Also, no one knows for sure where the tangram got its name. Theories range from it being named for Tan (a legendary Chinese scholar) to tanka (Chinese families who lived on riverboats) to Tang (a Chinese dynasty, 618–907 A.D.). I say again it doesn't matter. What does matter, is the name means a wonderful manipulative for problem–solving and for exploring geometric and other mathematical concepts. (Perhaps 'tan' means a TANgible asset in any classroom!)

JAN'S TAN BACKGROUND

I've tangrammed my way through many years of teaching. I tangrammed in a rural K–12 school (Hahira, Georgia), in a suburban junior high (Dayton, Ohio), and in an urban high school (Elmont, New York). I tangrammed as a substitute teacher (Cocoa Beach, Florida). I tangrammed in a pre–school and as a high school math chairperson (Santa Maria, California). And, in all my tangramming travels, I never met a kid who didn't like (or learn from) tangrams.

GETTING A TAN

You can buy or make a set of tans. (See pattern on page 45.)

KEEPING A TAN

Keep the tans handy. At first, let students just play with the pieces without instructions or guidance. Then, give them one or more worksheets from this book or do one of the teacher–directed lessons. By the way, sometimes a group of pages have been put in order (in TANdem?) by increasing difficulty. But, for the most part, you can skip around picking and choosing pages to suit your needs. (That's TANtamount to saying, "TAN to any page you wish!!")

JAN'S TANS

I've designed each worksheet so students of all ability levels can succeed. Each page features a "challenge" activity to extend the learning, to increase the time spent on each page, and to offer opportunities for creative and mathematical explorations.

Teacher pages are sprinkled throughout the book. (I never like it when they're all lumped in the beginning of a book.) There are ideas for bulleTAN boards, classroom projects, activity centers, holiday lessons, games, and family involvement. Tangram certificates are included to make it easy for you to recognize student achievement.

ANN'S, DAN'S AND FRAN'S TANS

And speaking of student achievement ... here's one last tip. Personalize your tangram lessons by having students make their own original designs and worksheets. Have students give them titles ... and sign their name.

OTHERS' TANS

Napoleon Bonaparte, Lewis Carroll, Edgar Allan Poe, John Quincy Adams, and Bruce Elbert Payne all enjoyed puzzling over their tangrams. You're in good company. Happy tangramming!

Jan Fair

P.S. 'Scuse the puns. I TAN't help it.

   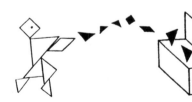

CLASSROOM LESSON: INTRODUCING THE TANGRAM

Give each student a set––and encourage them to experiment with the pieces. An overhead projector is nice for demonstration; you don't need any special overhead materials as a set of tangrams will project a shadowed outlined of each of the seven pieces.

MATERIALS:

Set of tangrams for yourself and each student (use the pattern on page 45 if you need to make them); overhead projector (optional)

GETTING TO KNOW THE PIECES

1. How many pieces are there in your puzzle? Sort the pieces into groups by shape. Sort by the number of sides. How many squares are there? Triangles? Which is the parallelogram? (Give students time to answer, then show the answers).
2. How are the various pieces alike? How are they different? Look at the square. What do you notice about its corners? Can you find a triangle which has a square corner? (Demonstrate by putting one piece on top of another).
3. Can you find a side of a triangle which is the same length as the sides of the square? (Move the two sides together to show they are the same).

MAKING NEW SHAPES

Use two of your pieces to make the square. Make a square from two other pieces. Make a parallelogram. Use two of the pieces to make a figure which is the same as another of your pieces. Use three pieces to make a house-shaped pentagon. Make a cat with your pieces.

OVERHEAD PROJECTOR FIGURES

1. (Turn off the overhead, make a shape with three of your pieces and then turn it back on). Make this same shape. (If students have trouble, move the shapes slightly apart to reveal the three pieces).
2. (Continue to make shapes and ask students to duplicate them. It will become more difficult as you increase the number of pieces you use. See the Task Cards on pages 35–41 for interesting figures for the students to solve).

THE SEVEN TANGRAM PIECES

Cover the shapes with your pieces.

CHALLENGE: Trace around each piece on the back of the paper. Give it to a friend to cover with his or her tangram pieces.

TANGRAM TREASURY, BOOK A, BY JAN FAIR, 1987 © CUISENAIRE COMPANY OF AMERICA, INC.

PARALLELOGRAM FLIP

Find the one tangram piece that fits.

TANGRAM TREASURY, BOOK A, BY JAN FAIR, 1987 © CUISENAIRE COMPANY OF AMERICA, INC.

FLIP AND FIT

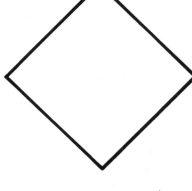

Match a piece to a shape. Then flip it over. Can you still make it fit? (Circle Yes or No)

1. (Yes)
   No

2. Yes
   No

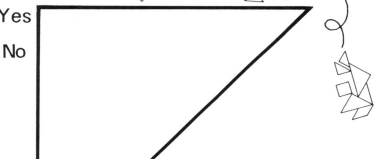

3. Yes
   No

4. Yes
   No

5. Yes
   No

7. Yes
   No

6. Yes
   No

CHALLENGE: Which pieces are the same?
Draw lines connecting them.

TANGRAM TREASURY, BOOK A, BY JAN FAIR, 1987 © CUISENAIRE COMPANY OF AMERICA, INC.

WHAT'S MY SHAPE?

Cover each shape with one tangram piece.

Ring its name.

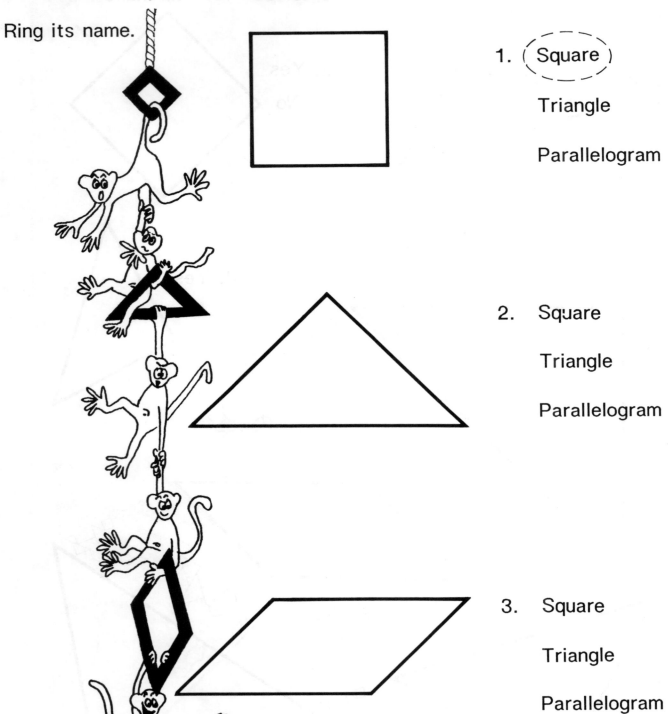

1. (Square)

    Triangle

    Parallelogram

2. Square

    Triangle

    Parallelogram

3. Square

    Triangle

    Parallelogram

CHALLENGE: Cover each shape with two tangram pieces.

TANGRAM TREASURY, BOOK A, BY JAN FAIR, 1987 © CUISENAIRE COMPANY OF AMERICA, INC.

FLYING SHAPES

Use the two large triangles to cover each shape.

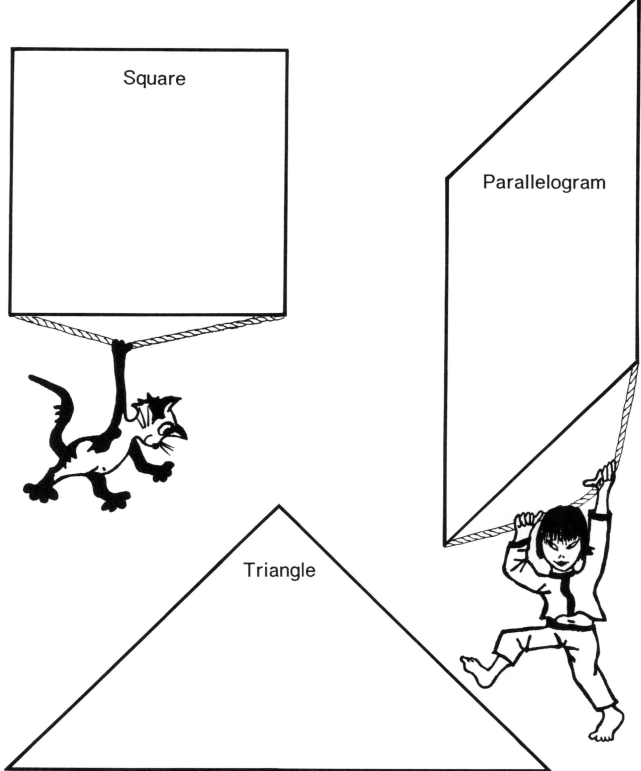

Square

Parallelogram

Triangle

CHALLENGE: In each figure, draw a line to show the two triangles. Color them in different colors.

TANGRAM TREASURY, BOOK A, BY JAN FAIR, 1987 © CUISENAIRE COMPANY OF AMERICA, INC.

A DOG–GONE PUZZLE

Cover the dog using all seven tangram pieces.

CHALLENGE: Make something else with your pieces. Ask a friend to guess what it is.

TANGRAM TREASURY, BOOK A, BY JAN FAIR, 1987 © CUISENAIRE COMPANY OF AMERICA, INC.

TANGRAM T-PARTY

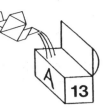

Cover the "T" using all seven tangram pieces.

CHALLENGE: Make another letter of the alphabet.

TANGRAM TREASURY, BOOK A, BY JAN FAIR, 1987 © CUISENAIRE COMPANY OF AMERICA, INC.

COVERED BABY CARRIAGE

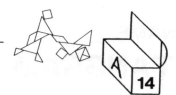

Cover the carriage with your tangram pieces.

CHALLENGE: Draw someone or something in the carriage. Color the picture.

TANGRAM TREASURY, BOOK A, BY JAN FAIR, 1987 © CUISENAIRE COMPANY OF AMERICA, INC.

NAME

TANGRAM LADY

Cover this figure with all
seven tangram pieces.

CHALLENGE: Use your
tangram pieces to make
this lady again on a
different piece of paper.

TANGRAM TREASURY, BOOK A, BY JAN FAIR, 1987 © CUISENAIRE COMPANY OF AMERICA, INC.

SPIDER ART

Find the hidden tangram pieces. Cover them.

CHALLENGE: Color the picture.

TANGRAM TREASURY, BOOK A, BY JAN FAIR, 1987 © CUISENAIRE COMPANY OF AMERICA, INC.

hid–TAN PICTURE

Find the hidden tangram pieces. What did you make? Shade or color it.

CHALLENGE: Make a hid–TAN picture and have a friend solve it.

TANGRAM TREASURY, BOOK A, BY JAN FAIR, 1987 © CUISENAIRE COMPANY OF AMERICA, INC.

FOR: 2 to 4 players

MATERIALS:

one set of tangrams for each player

one worksheet (pages 12 through 17) or one task card <u>with</u> lines (A–35 through A–40), reproduced for each player as a gamesheet

spinners (page 19)

NOTE:

You might want to do the worksheet before using it as a gamesheet.

DIRECTIONS:

1. Players each spin the number spinner. Highest number goes first.
2. On your turn, spin the shape spinner. If you have that piece in your tangram set, use it to cover the matching shape on your gamesheet.
3. The winner is the first player to cover the entire picture on his or her gamesheet.

CHALLENGE:

To make the game more difficult, use one of the following as gamesheets: pages 32 and 33 or the task cards A–35 through A–40, which have <u>no</u> lines to show individual pieces. On each turn, after a player has positioned a piece, it cannot be moved until that player has spun the same piece again. Moving a piece to a different position counts as a turn.

FOR: 2 to 4 players

MATERIALS:

one set of tangrams for each player

spinners (page 19)

paper and pencil for scoring

NOTE:

Before playing this game, you might want to practice making squares, triangles, and parallelograms by using one, two, three, and then four tangram pieces.

DIRECTIONS:

1. On each round of play, a player spins the shape spinner to determine the shape to be used that round. Then, each player in turn spins a number and uses that number of pieces to make the shape. The shape made <u>does not</u> have to be the same size as the shape spun. A player's score for the round is the number spun only if he or she is able to make the shape. (If 3 or 4 people are playing, each person can first spin a number and then everyone can work at the same time to make the shape.)
2. A game can end after any round. Players total their scores. The winner is the player with the highest total.

CHALLENGE:

To make the game more difficult, when players spin a "1", they must spin again and use "one more" than the number of pieces spun the second time.

SPINNERS

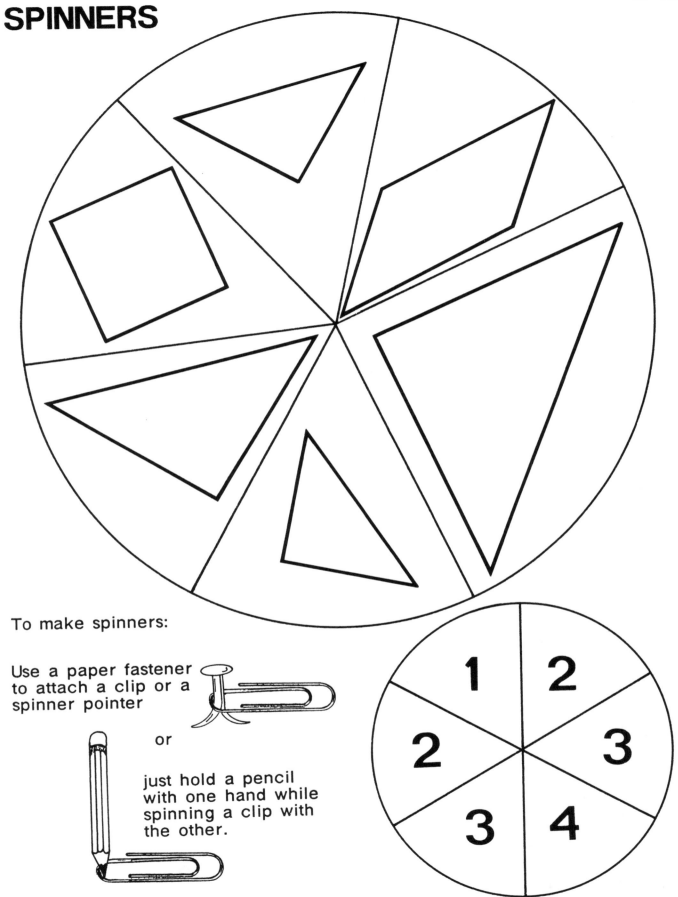

To make spinners:

Use a paper fastener to attach a clip or a spinner pointer

or

just hold a pencil with one hand while spinning a clip with the other.

TANGRAM TREASURY, BOOK A, BY JAN FAIR, 1987 © CUISENAIRE COMPANY OF AMERICA, INC.

NAME

MODERN ART

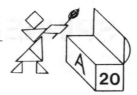

Trace around each of your seven tangram pieces. Color.

CHALLENGE: Use your modern art as a puzzle. Ask someone to cover with pieces that match.

TANGRAM TREASURY, BOOK A, BY JAN FAIR, 1987 © CUISENAIRE COMPANY OF AMERICA, INC.

NAME

KITE FLYING SHAPES

Cover with tangram pieces.
Trace around them.

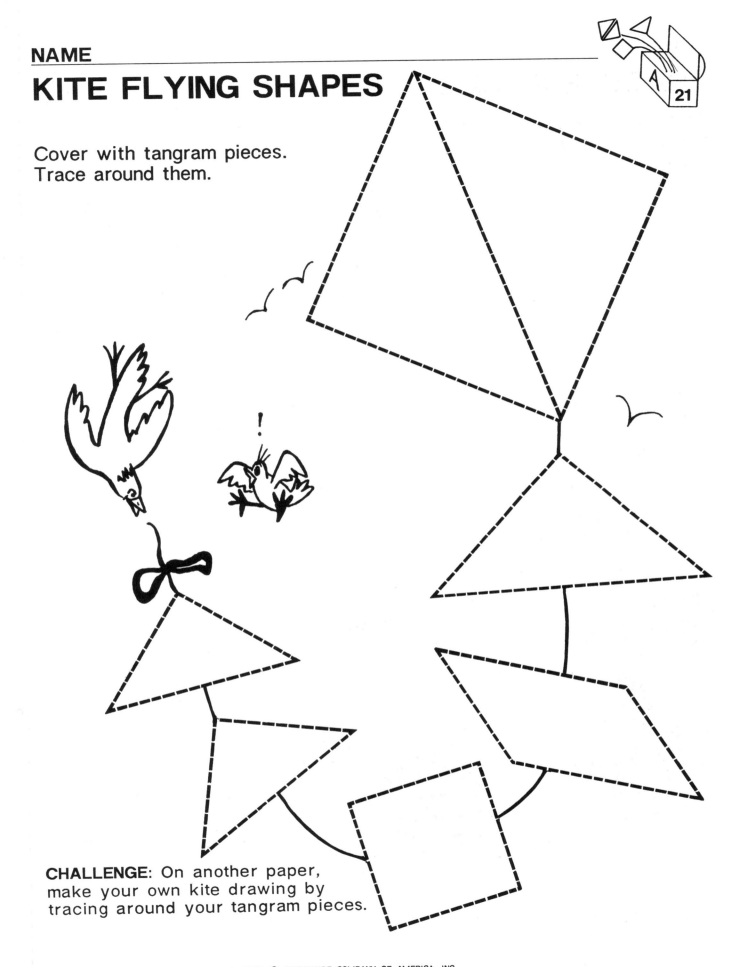

CHALLENGE: On another paper,
make your own kite drawing by
tracing around your tangram pieces.

TRACE A RACER

Cover with the seven tangram
pieces. Trace around each.

CHALLENGE: On another paper, use your pieces to make this racer.
Trace to record.

TANGRAM TREASURY, BOOK A, BY JAN FAIR, 1987 © CUISENAIRE COMPANY OF AMERICA, INC.

DON'T FLIP! ROTATE!

Cover the shape with one piece. Then rotate the piece and trace around it.

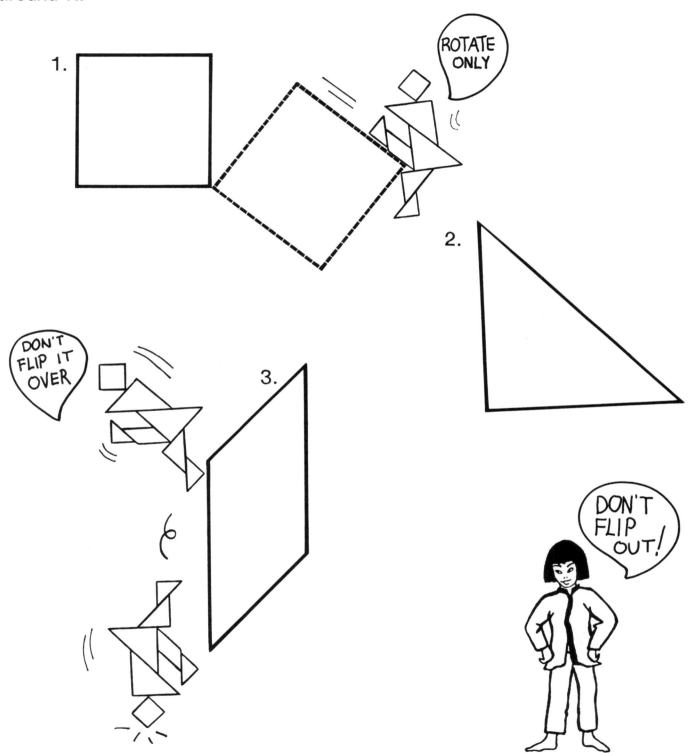

CHALLENGE: On another paper, make a design by rotating and tracing pieces.

TANGRAM TREASURY, BOOK A, BY JAN FAIR, 1987 © CUISENAIRE COMPANY OF AMERICA, INC.

TRIANGLE PINWHEEL

Cover with a triangle that matches. Trace around each.

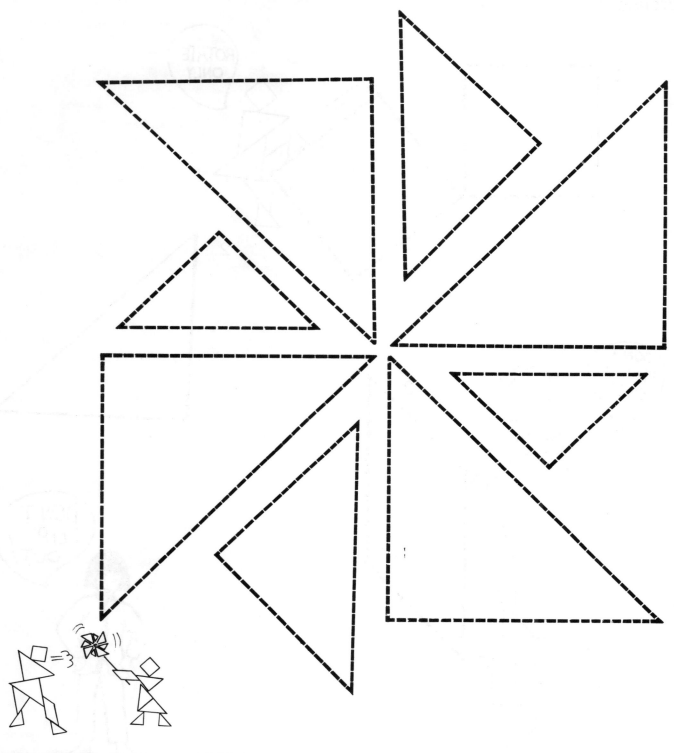

CHALLENGE: Trace a pinwheel on another paper. Color it.

TANGRAM TREASURY, BOOK A, BY JAN FAIR, 1987 © CUISENAIRE COMPANY OF AMERICA, INC.

FLOWER ROTATIONS

Match your piece to the shape. Rotate it so a point is on the circle.
Trace around it. Rotate and trace around it until you have a flower.

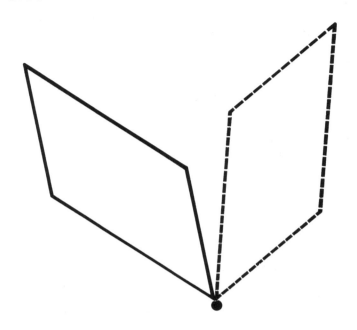

CHALLENGE: Make a flower by tracing around <u>triangle</u> pieces.

PROJECT: TRANSFORMATIONAL TANGRAM FLOWERS

This is a "geometric" arts and crafts project the entire class can do. Students work at their individual ability levels to make flowers from tangram pieces. The methods for making a flower focus on geometric concepts including transformational geometry which is concerned with the movement of geometric figures. The project can consist of one simple flower or a big complicated flower arrangement. And, it can all culminate into a spectacular bulletin board!

MATERIALS:

Tangrams; construction paper; markers or crayons; scissors; glue; TANplate (optional)

NOTES:

Before doing this project, you might want students to do one or more of the following worksheets: DON'T FLIP! ROTATE! (page 23), TRIANGLE PINWHEEL (page 24) and FLOWER ROTATIONS (page 25).

PREPARATION (optional):

Make a few samples for students to see.

DIRECTIONS:

Students use their tangram pieces as a pattern to trace around several times in a configuration that looks like a flower. Or, they can trace and cut out tangram shapes from construction paper--then arrange and attach them to paper of a contrasting color. Students can work alone or in groups to arrange several flowers in a small bouquet.

USE ONLY ONE PIECE

Make a flower using only triangles (see figures A, E and F on the next page), or only parallelograms (figure B).

USE DIFFERENT PIECES

Make a flower using all five of the tangram shapes (figure D).

SLIDE A PIECE

Trace around one piece. "Slide" it to a new location. (Remind students that this motion changes only the figure's position, not its size or shape). When you "slide" a figure, be careful not to turn it. Sides that were horizontal, vertical, or slanted are still horizontal, vertical, or slanted in the same direction (figure C).

ROTATE A PIECE

Trace around one piece. "Rotate" it, and trace again. For example, choose a triangle. Rotate it so that one of its angles stays on some point which will then become the center of the flower (figures A and E).

REFLECTION FLOWERS

Trace around the parallelogram piece. "Flip" it over and trace again. (See figure B).

SYMMETRICAL FLOWERS

Make a flower such that it is symmetrical around a line (figures B,C, and E). Use a mirror to prove it is symmetrical. As a challenge, make a flower bouquet which is symmetrical around a vertical line (figure E).

TANGRAM FLOWERS

A

B

C

D

E

F

SQUARES IN SQUARES

How many tangram squares will cover each square?

1. _____

2. _____

3. _____

CHALLENGE: Make a square with nine tangram squares.

SHOW YOUR ANSWER BY TRACING

WHAT A SQUARE!

TANGRAM TREASURY, BOOK A, BY JAN FAIR, 1987 © CUISENAIRE COMPANY OF AMERICA, INC.

TRIANGLES IN TRIANGLES

How many small triangles will cover each triangle?

1. _____

2. _____

3. _____

SHOW YOUR ANSWER BY TRACING

4. _____

CHALLENGE: Make a triangle with more than eight small triangle pieces.

TANGRAM TREASURY, BOOK A, BY JAN FAIR, 1987 © CUISENAIRE COMPANY OF AMERICA, INC.

A FISHY COVER–UP

Cover the fish shape with several tangram pieces.

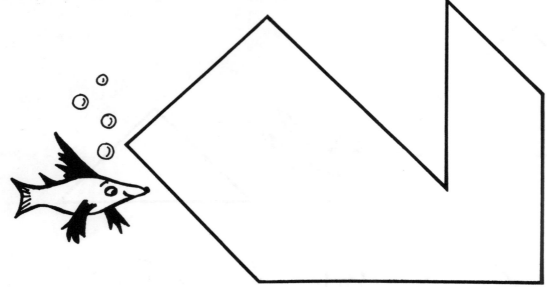

WRITE A NUMBER HERE

1. Tell what pieces you used.

_____ large triangle

_____ medium triangle

_____ small triangle

_____ parallelogram

_____ square

2. Cover it a different way and tell what pieces you used.

_____ large triangle

_____ medium triangle

_____ small triangle

_____ parallelogram

_____ square

CHALLENGE: Make a different fish. Give it to a friend to cover in two ways.

TANGRAM TREASURY, BOOK A, BY JAN FAIR, 1987 © CUISENAIRE COMPANY OF AMERICA, INC.

TWO-WAY TRIANGLES

Cover the triangle with pieces from your tangram set.

1. Tell what pieces you used.

_____ large triangle

_____ medium triangle

_____ small triangle

_____ parallelogram

_____ square

WRITE A NUMBER HERE

2. Cover it a different way.
   Tell what pieces you used.

_____ large triangle

_____ medium triangle

_____ small triangle

_____ parallelogram

_____ square

CHALLENGE: Show one of the ways you covered the triangle above by tracing around the pieces.

COVER IT WELL

Use all seven pieces to cover the figure.

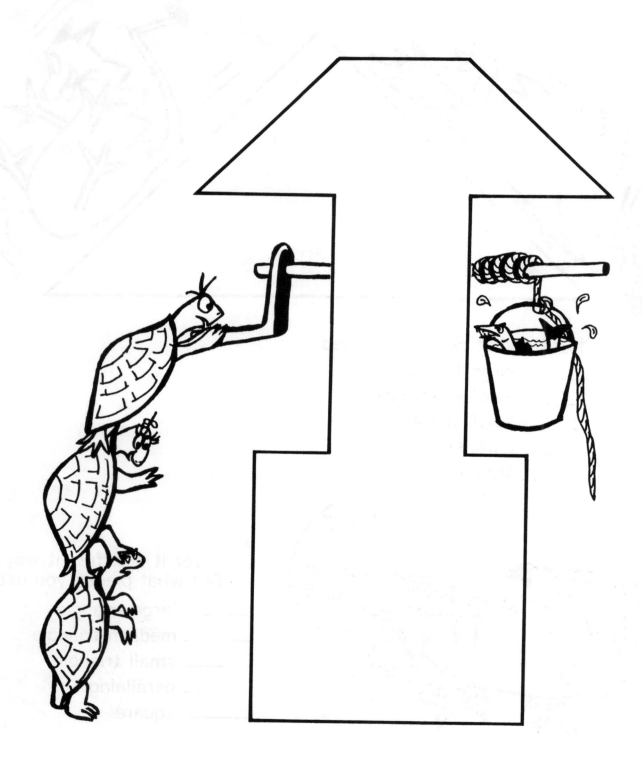

CHALLENGE: Turn this paper over. Can you make this figure again?

NAME

FUNNY BUNNY

Cover the figure
using all seven
tangram pieces.

CHALLENGE: Use your pieces to make a different rabbit.

TANGRAM TREASURY, BOOK A, BY JAN FAIR, 1987 © CUISENAIRE COMPANY OF AMERICA, INC.

LEARNING CENTERS

Tangram activities lend themselves easily for use in learning centers. Students can work independently on most activities. And, as students progress, they will get immediate feedback as to how they are doing since most activities are self-checking. Below are some ideas for ways this book can help in your learning centers.

SELF-CHECKING TASK CARDS

MATERIALS:

Task cards (pages 35–41); tangrams

PREPARATION:

1. Make copies of the task cards, cut them apart and store in plastic "sleeves". They can also be pasted or stapled to stiff paper and laminated or covered with contact paper.
2. Post a sign at the station which says, "Choose any card and cover the shape with tangram pieces."

DIRECTIONS:

You might want to have students first do the side of several task cards showing all of the lines which divide the shape into pieces. Then have students pick and do task cards showing the shape's outline only. Have the more advanced student make the figure 'next' to the task card rather than 'on' the task cards.

WORKSHEET "INSTANT" CENTERS

Although the task cards work best, you can also use any worksheet from this book in a learning center.

INDEPENDENT ART PROJECTS

MATERIALS:

Tangrams; construction paper; markers or crayons; scissors; glue; TANplate (optional)

PREPARATION (optional):

Make a few samples for students to refer to at the learning center.

DIRECTIONS:

MODERN ART (page 20) and TRANSFORMATIONAL TANGRAM FLOWERS (pages 26–27) can be done by students at a learning center. They can also make "farm" figures for a BulleTAN Board (pages 41–42).

CAT

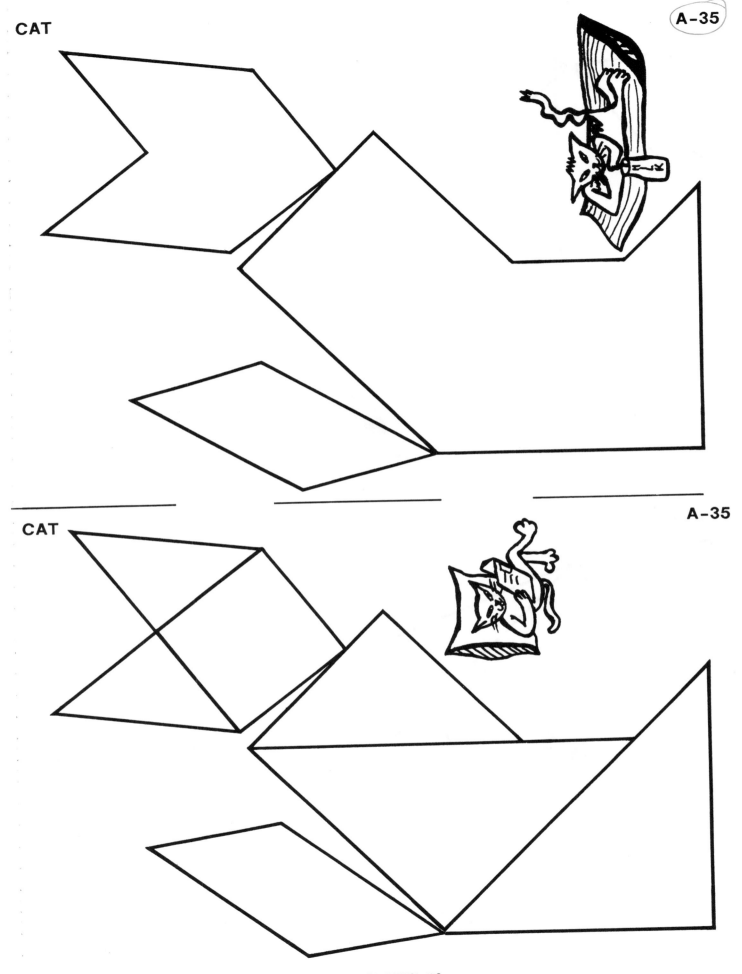

CAT

TANGRAM TREASURY, BOOK A, BY JAN FAIR, 1987 © CUISENAIRE COMPANY OF AMERICA, INC.

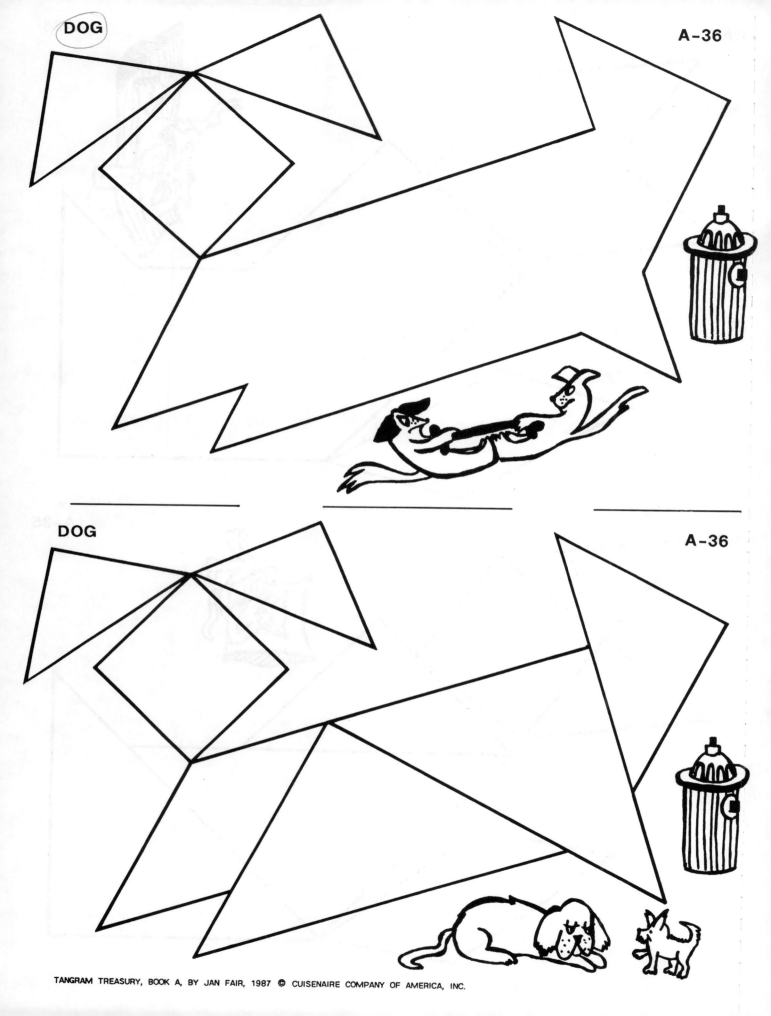

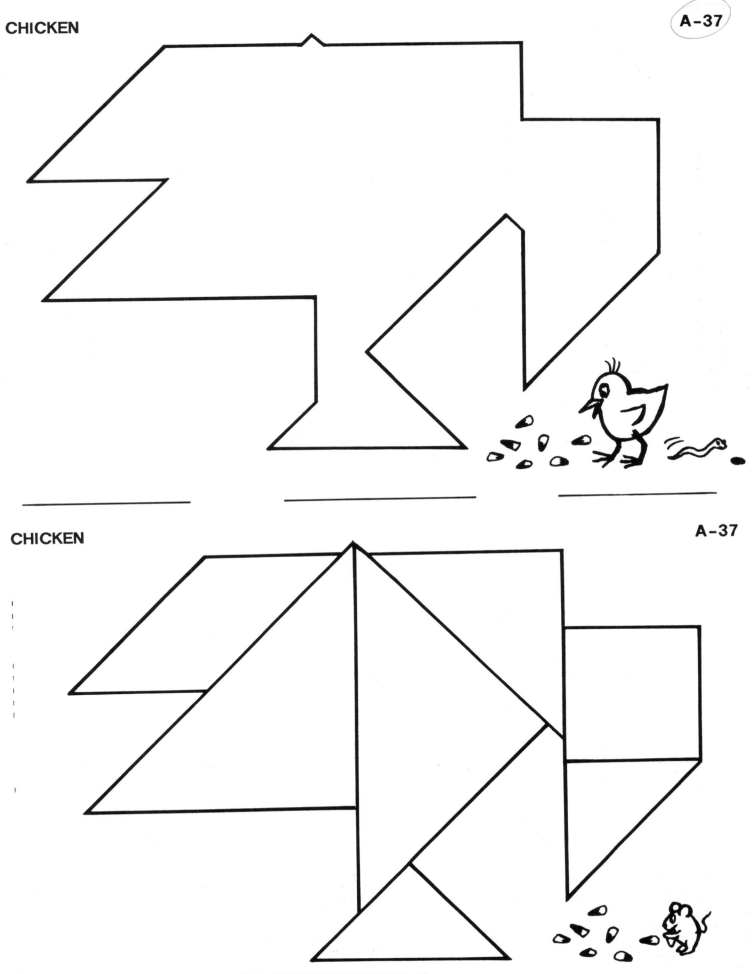

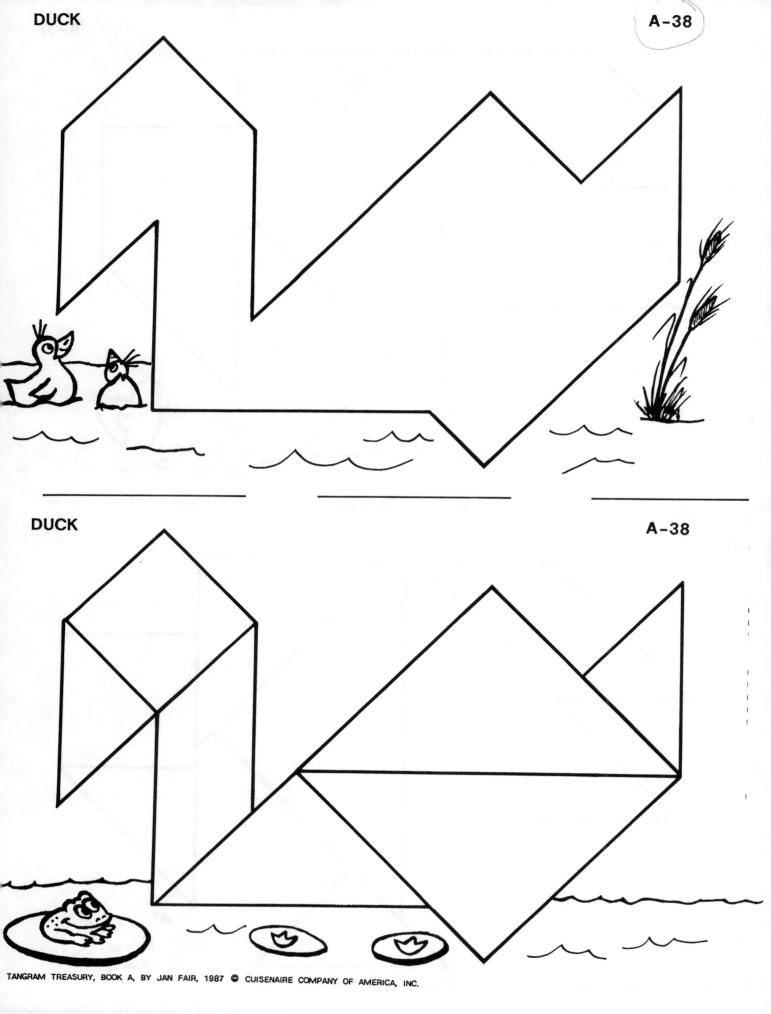

DUCK

A-38

TREES

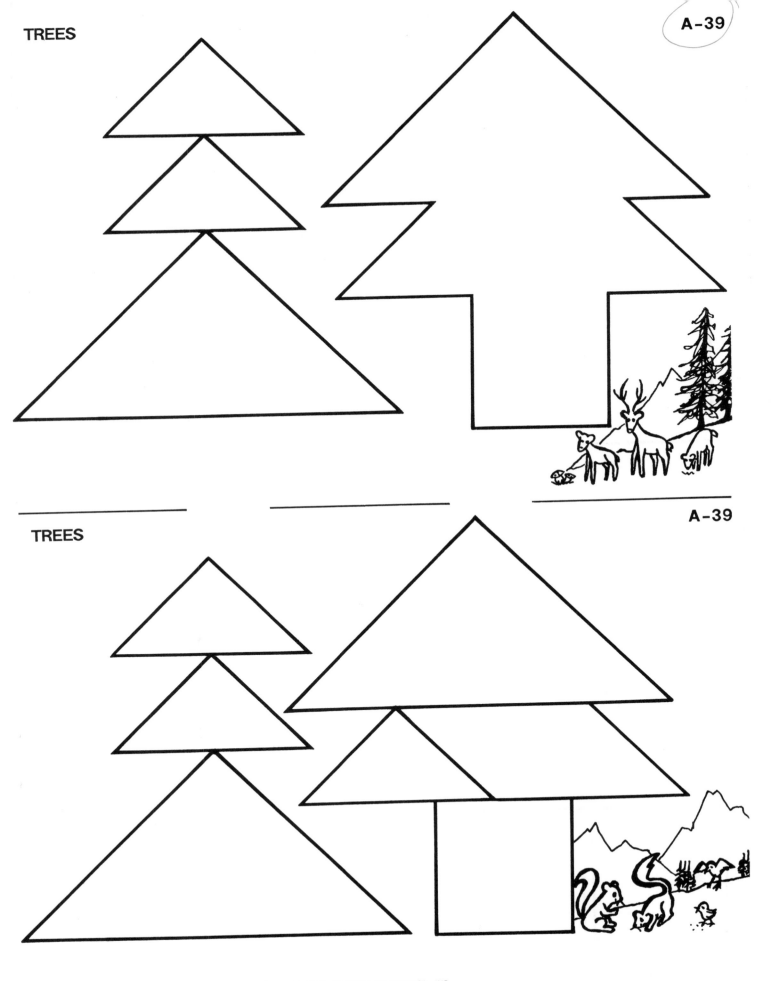

TREES

TANGRAM TREASURY, BOOK A, BY JAN FAIR, 1987 © CUISENAIRE COMPANY OF AMERICA, INC.

BARN

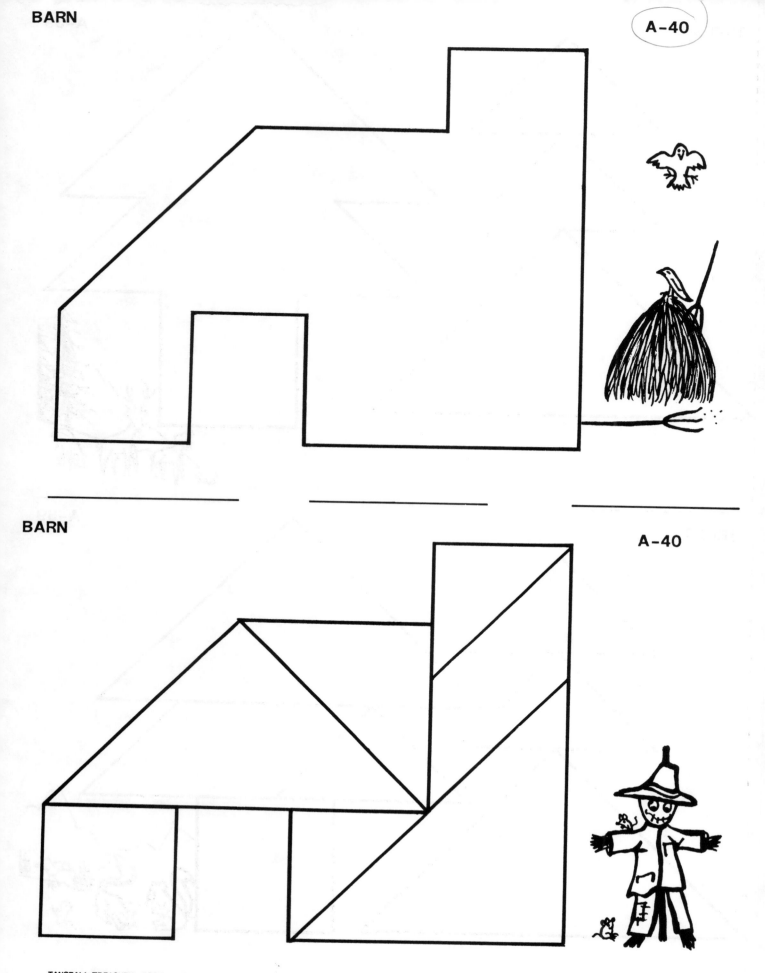

BARN

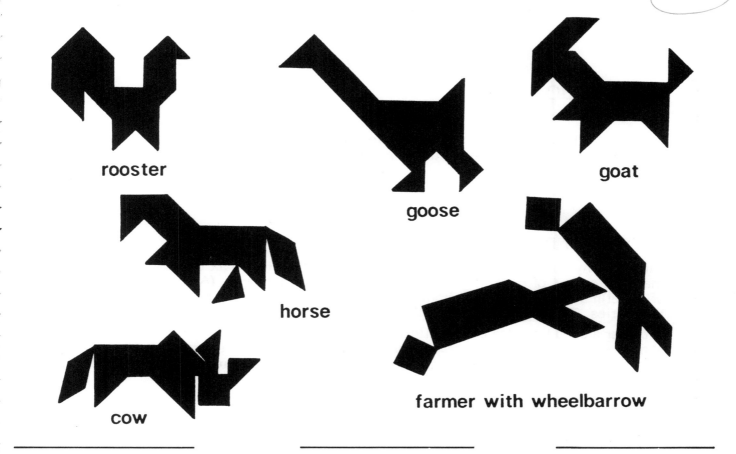

rooster

goose

goat

horse

cow

farmer with wheelbarrow

TANGRAM FARM

A-41

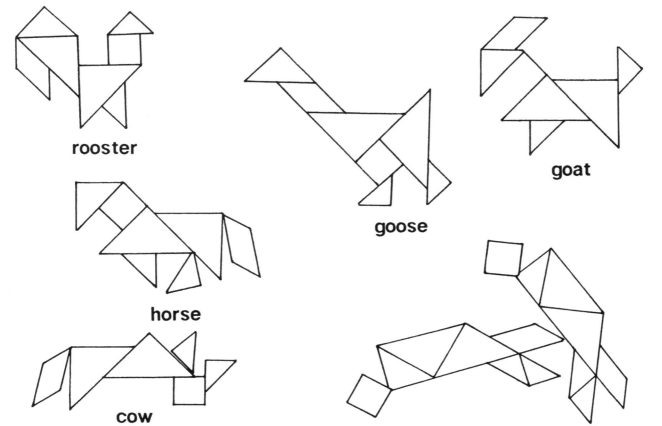

rooster

goose

goat

horse

cow

farmer with wheelbarrow

bulleTAN BOARDS

Tangram activities can result in some lively and attractive bulletin boards. And, when students are involved in making them, it not only saves you time, it stimulates learning as well.

LETTERS AND BORDERS

Although any lettering will do on a bulletin board, you might want to have students make letters out of tangram shapes cut from construction paper. (A complete set of bulletin board letters made from tangrams can be found in the second book in this series, Tangram Treasury, Book B.)

A border around the edge can highlight a bulletin board. Have students cut out tangram shapes and attach them with staples or pins. Depending on the season or holiday, they might want to put shapes together to form appropriate figures such as snowflakes, flowers, or rabbits.

TANGRAM FARM

MATERIALS:
Task cards (pages 35–41); tangrams; construction paper; markers or crayons; glue; TANplate (optional)

DIRECTIONS:
1. Cover the bulletin board with paper. If you like, make a title and a border. (See LETTERS AND BORDERS above.)
2. Have students work alone or in groups to make 'farm' figures for the board by following the patterns found on the task cards. They can use their tangram pieces as a pattern to trace, and then color in the figures. Or, they can trace and then cut out tangram shapes from construction paper and arrange and attach them to the bulletin board––or paste them on another paper before posting them. (See page 43 for one possible way to arrange the figures.)

TANGRAM ART

DIRECTIONS:
1. For an "instant" bulletin board, just post students' work after they've done worksheets such as hid–TAN PICTURE (page 17), MODERN ART (page 20) and FLOWER ROTATIONS (page 25).
2. Another, more spectacular bulletin board can result from your students doing, as a class project, TRANSFORMATIONAL TANGRAM FLOWERS (pages 26–27).

TANGRAM TREASURY, BOOK A, BY JAN FAIR, 1987 © CUISENAIRE COMPANY OF AMERICA, INC.

TanGRam FaRm

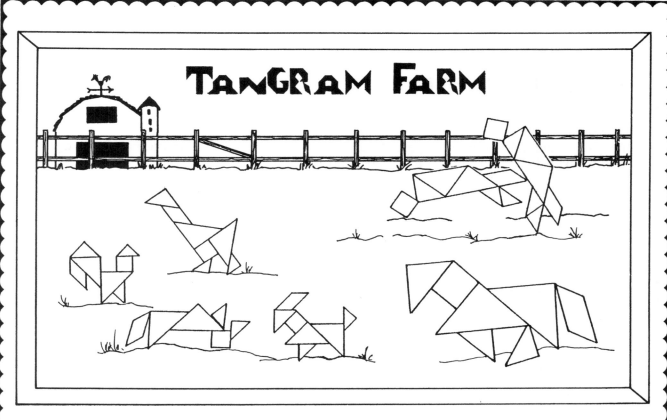

TART TANGRAM

FAMILY INVOLVEMENT

Tangram activities provide an especially nice way to encourage communication between home and school. They offer parents an intriguing way to share in their child's learning and to experience, firsthand, an interesting mathematics manipulative. Below you will find some ideas for ways to use this book to help encourage family involvement.

MATERIALS:
TAN-gram and puzzle pieces (page 45) or the blank TAN-gram (page 46) if you want to write your own message to parents; one or more worksheets or student-made pages.

REPEATED SUCCESS

A good way to communicate with parents is to have students complete an activity successfully in class ...and then repeat that success at home with their parents. Have students take home a duplicate copy of one or more worksheets which they have completed successfully, along with a TAN-gram message from you. Encourage students to show their parents how they did the activity and talk about what they've learned.

A STUDENT-MADE ACTIVITY

Students can make a tangram activity for their family to do. In class, have them use their puzzle pieces to make a figure such as an animal, letter of the alphabet, or some object. Then, give students a piece of paper and ask them to trace around each shape or around the outside edge of the figure only. Send home a TAN-gram along with this paper and have students challenge a family member to cover the picture with the seven tangram pieces.

Other ideas for student-made activities are found in the CHALLENGE activities at the bottom of the following worksheets: hid-TAN PICTURE (page 17), TANGRAM MODERN ART (page 20), and FUNNY BUNNY (page 33).

FAMILY GAMES

Students can play one or more tangram games with their family. Have them play the games at school before taking them home. Afterwards, encourage students to share with the class their experiences playing the games at home.

To play a game at home, students will need a copy of the TAN-gram (page 45), the game instructions (page 18), and the game spinners (page 19).

CHALLENGE ACTIVITIES

After students have completed the main part of a worksheet, assign the CHALLENGE activity at the bottom of the worksheet to be done at home with their families--who may be surprised to find they are challenged by these activities as well!

TANGRAM TREASURY, BOOK A, BY JAN FAIR, 1987 © CUISENAIRE COMPANY OF AMERICA, INC.

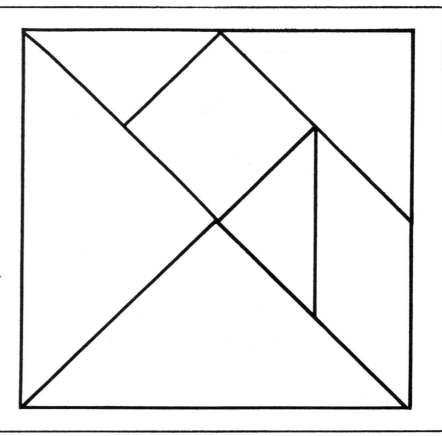

Tan-Gram

wes-tan union

DATE:

TO: Family

MESSAGE: In class we've been using tangrams to explore geometric and other mathematical concepts. This seven-piece puzzle which originates from China can be intriguing to young and old alike. Use them to form a variety of geometric shapes, letters of the alphabet, animals, people and objects.

SENDER'S NAME:

PAGES ATTACHED:

TANGRAM PATTERN

To make a set of tangrams, cut out the seven pieces and store them in an envelope or plastic bag.

To make a more durable set, use this as a pattern to make pieces from heavy paper, cardboard, wood, or even floor tiles. (Warm tiles in an oven for easier cutting.)

Use solid colors.

TANGRAM TREASURY, BOOK A, BY JAN FAIR, 1987 © CUISENAIRE COMPANY OF AMERICA, INC.

Tangram
Certificate

NAME _____

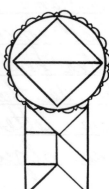

_____

_____

_____

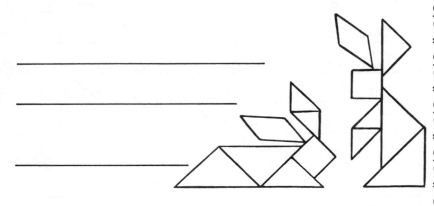

![wes-tan union] **wes-tan union**		**Tan-Gram**
		DATE:
TO: _____		
_____		
MESSAGE: _____		
_____		
_____		
_____		
_____		
SENDER'S NAME:		

SELECTED ANSWERS

FLIP AND FIT

Match a piece to a shape. Then flip it over. Can you still make it fit? (Circle Yes or No)

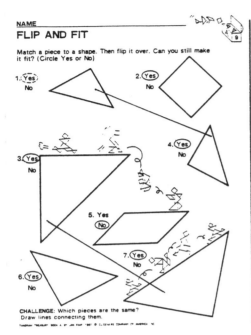

1. (Yes) No
2. (Yes) No
3. (Yes) No
4. (Yes) No
5. Yes (No)
6. (Yes) No
7. (Yes) No

CHALLENGE: Which pieces are the same? Draw lines connecting them.

WHAT'S MY SHAPE?

Cover each shape with one tangram piece.

Ring its name.

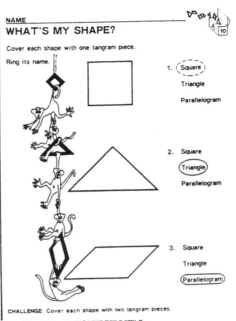

1. (Square) Triangle Parallelogram
2. Square (Triangle) Parallelogram
3. Square Triangle (Parallelogram)

CHALLENGE: Cover each shape with two tangram pieces.

FLYING SHAPES

Use the two large triangles to cover each shape.

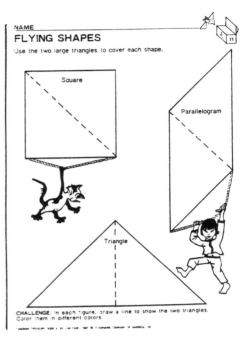

Square

Parallelogram

Triangle

CHALLENGE: In each figure, draw a line to show the two triangles. Color them in different colors.

SPIDER ART

Find the hidden tangram pieces. Cover them.

CHALLENGE: Color the picture.

hid-TAN PICTURE

Find the hidden tangram pieces. What did you make? Shade or color it.

CHALLENGE: Make a hid-TAN picture and have a friend solve it.

DON'T FLIP! ROTATE!

Cover the shape with one piece. Then rotate the piece and trace around it.

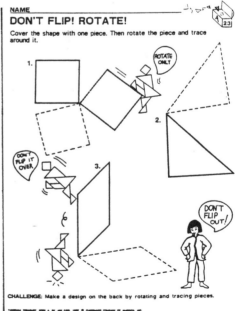

1.
2.
3.

ROTATE ONLY

DON'T FLIP IT OVER

DON'T FLIP OUT!

CHALLENGE: Make a design on the back by rotating and tracing pieces.

SELECTED ANSWERS (continued)

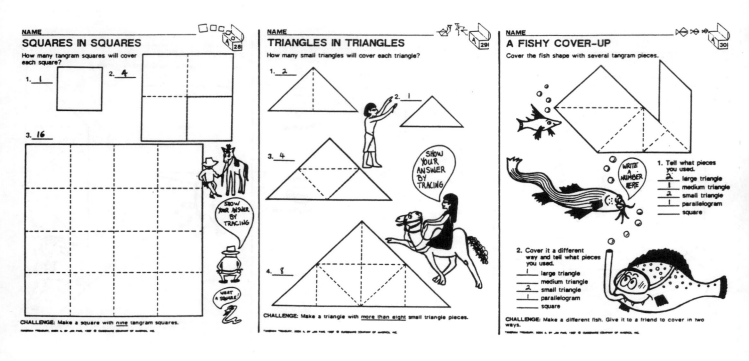

NAME _____

SQUARES IN SQUARES

How many tangram squares will cover each square?

1. 1
2. 4
3. 16

CHALLENGE: Make a square with nine tangram squares.

NAME _____

TRIANGLES IN TRIANGLES

How many small triangles will cover each triangle?

1. 2
2. 1
3. 4
4. 8

SHOW YOUR ANSWER BY TRACING

CHALLENGE: Make a triangle with more than eight small triangle pieces.

NAME _____

A FISHY COVER-UP

Cover the fish shape with several tangram pieces.

WRITE A NUMBER HERE

1. Tell what pieces you used.
   2 large triangle
   1 medium triangle
   2 small triangle
   1 parallelogram
   ___ square

2. Cover it a different way and tell what pieces you used.
   1 large triangle
   2 medium triangle
   2 small triangle
   1 parallelogram
   ___ square

CHALLENGE: Make a different fish. Give it to a friend to cover in two ways.

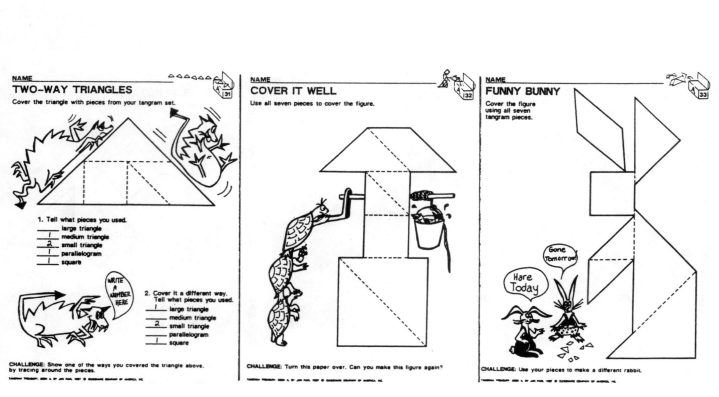

NAME _____

TWO-WAY TRIANGLES

Cover the triangle with pieces from your tangram set.

1. Tell what pieces you used.
   1 large triangle
   ___ medium triangle
   2 small triangle
   1 parallelogram
   1 square

WRITE A NUMBER HERE

2. Cover it a different way. Tell what pieces you used.
   1 large triangle
   ___ medium triangle
   2 small triangle
   ___ parallelogram
   1 square

CHALLENGE: Show one of the ways you covered the triangle above by tracing around the pieces.

NAME _____

COVER IT WELL

Use all seven pieces to cover the figure.

CHALLENGE: Turn this paper over. Can you make this figure again?

NAME _____

FUNNY BUNNY

Cover the figure using all seven tangram pieces.

Hare Today

Gone Tomorrow!

CHALLENGE: Use your pieces to make a different rabbit.

TANGRAM TREASURY, BOOK A, BY JAN FAIR, 1987 © CUISENAIRE COMPANY OF AMERICA, INC.